CLOUDS

&

WEATHER PHENOMENA

Fig. 1. Long band of cirrus

CLOUDS

&

WEATHER PHENOMENA

C. J. P. CAVE, M.A.

President of the Royal Meteorological Society
1913–1915 & 1924–1926

Late Captain, Meteorological Section,
Royal Engineers

A NEW EDITION
Revised and with many new photographs

AT THE UNIVERSITY PRESS

To

M. C.

CAMBRIDGE
UNIVERSITY PRESS

University Printing House, Cambridge CB2 8BS, United Kingdom

Cambridge University Press is part of the University of Cambridge.

It furthers the University's mission by disseminating knowledge in the pursuit of education, learning and research at the highest international levels of excellence.

www.cambridge.org
Information on this title: www.cambridge.org/9781107504868

First edition 1926
Reprinted twice 1926
Second edition 1943
First published 1943
First paperback edition 2015

A catalogue record for this publication is available from the British Library

ISBN 978-1-107-50486-8 Paperback

CONTENTS

PREFACE

A CERTAIN knowledge about clouds, rainbows, halos, mirages and other phenomena of the atmosphere is useful to many who, while not wishing to make a deep study of meteorological matters, yet take an interest in the changing pageant of the sky. The usual books on the subject are sometimes too technical or contain much information on meteorological instruments, the principles of forecasting from weather maps, and other matters, which, though extremely interesting in themselves, are not wanted by those who depict or watch the skies from a less scientific standpoint. It is for these that this book is written.

The photographs of clouds which appear in the volume were mostly taken with a view of getting as much detail as possible of the cloud structure, and not with a view of making artistic pictures; they are what is known to photographers as contrasty, and may be too contrasty for pictures; some of them have been dubbed too theatrical; in many of them the blue of the sky is almost black; they are reproduced here to illustrate the various forms of cloud structure. For the information of those who may wish to take cloud photographs it may be mentioned that most of those reproduced here were taken on backed panchromatic plates through a deep yellow or red screen; one was taken through an infra-red screen.

My thanks are due to Professor D. Brunt, F.R.S., for reading the manuscript and for some valuable suggestions; also to Lieut.-Colonel E. Gold, D.S.O., F.R.S., for information about condensation trails.

C. J. P. C.

October, 1943

ILLUSTRATIONS

CLOUDS

A KNOWLEDGE of the different forms of clouds may be very useful to those who are interested in the changes of weather, so frequent in our climate, and such knowledge may be extremely useful to those who navigate the seas and still more to those who navigate the air. Such a knowledge is easily obtained, for though meteorologists have subdivided the forms of clouds into a great number of varieties yet the main groups are few in number and can readily be learnt by anyone who takes the trouble to observe them. Luke Howard (1772–1864) published a classification of clouds quite early in the nineteenth century, and with certain modifications the names he gave have survived to the present day. He gave Latin names to the different forms of clouds, but these Latin names have become part of our own language, and cirrus, cumulus and nimbus are given as English words in English dictionaries. Compound names like cirrocumulus originally indicated that the cloud was supposed to be a sort of cross between cirrus and cumulus; with our present-day knowledge we know that this is not the case, and these compound names must be looked on as just names and nothing more.

Before coming to the actual classification of the clouds it may be useful to realise that clouds may be divided into two main groups, which we may call sheet clouds and heap clouds. The sheet clouds are arranged in horizontal layers which as a rule are of no great vertical thickness. They may form extensive sheets covering the whole sky, or they may be broken up, so that patches of blue sky are seen through the sheet, or again they may be reduced to a few detached clouds, which, however, are all floating at some definite height above the earth. There may be several layers of cloud at different heights and the upper may be seen through gaps in the lower (Fig. 2). The sheet clouds are sometimes marked by beautiful wave or ripple patterns; sometimes they consist of long parallel waves which owing to the effect of perspective seem to radiate from a point on the horizon.

A cloud sheet will form at any layer of the atmosphere where conditions of temperature and moisture cause condensation to take place, and there may be more than one layer where such conditions occur.

Heap clouds are piled-up clouds whose thickness may be as great or greater than their vertical extent; they may be many thousands of feet in thickness. They are formed when currents of air rise in the atmosphere; the air in the process of rising expands, in expanding it cools until a point is reached where it grows so cold that part of the moisture it contains condenses and cloud is formed.

With these facts in mind we may now consider the different varieties of cloud, and the names that have been given to them.

I. Clouds in horizontal sheets, sometimes covering the whole sky, sometimes more or less broken up, and sometimes reduced to a few isolated patches.

1. Cirrus (from the Latin, a curl or tuft of hair). Clouds composed of threads or fibres, sometimes in isolated wisps or tufts (Figs. 3, 4), sometimes in compact masses (Fig. 5), or, it may be, arranged in long lines across the sky (Frontispiece and Fig. 6). Some varieties are known by the name of mares' tails. Often the streaks of cirrus have their ends turned up into little tufts (Figs. 2, 7, 8); sometimes the tuft is the principal part of the cloud, giving rise to the form known as tailed cirrus (Fig. 9). Cirrus may cover the whole sky, sometimes in detached masses, sometimes in a continuous sheet. In the latter case it is known as cirrostratus (Fig. 11), and may be made up of tangled fibres or threads, or it may be a structureless sheet giving a white milky look to a part or the whole of the sky. The sun can always be

seen through a sheet of cirrostratus. Sometimes the sheet may be so thin as to be invisible, merely making the blue of the sky a little paler, but such a sheet may be recognisable by the halos that accompany it. (See under 'Halos'.)

These clouds, cirrus and cirrostratus, are composed of minute ice crystals in the higher part of the atmosphere; practically all other forms of clouds are composed of minute drops of water. Cirrus clouds are usually at a height of 20,000 to 30,000 feet in our latitudes, but near the equator they may at times be twice that height; they may be seen floating above the tops of high mountains; they often move from a different direction to the surface wind and the lower clouds, as they may be in a different wind current. They have a tendency to move outwards from areas where the barometer is low, and when seen moving from the west or north-west with a southerly (south-east to south-west) surface wind it is usually a sign of bad weather advancing from the west. The frayed-out fibres of cirrus clouds are popularly supposed to mean wind, and a windy sky in popular language is synonymous with such clouds, but the connection is not a real one; we often see frayed-out streaks and masses of cirrus in very quiet weather.

2. Cirrocumulus consists of small cloudlets, or waves and ripples all of which are more or less mingled with cirrus threads (Fig. 13); the cloudlets are pure white and show no shadows; in the absence of intermingled cirrus and the presence of shadow a cloud should not be classed as cirrocumulus but as the next variety, altocumulus. The cloudlets of cirrocumulus are generally arranged in some sort of order, that is, in bands or waves (Figs. 12, 14, 15), and very often the wave structure is well marked; there may often be two or more sets of waves crossing each other at various angles (Fig. 13). This variety of design produces the most beautiful skyscapes, and gives rise to the so-called mackerel skies. Sometimes long bands of cirrus and cirrocumulus clouds trail across the sky, usually from a northerly or north-westerly point; these polar bands, as they are termed, may

be a hundred miles long or more, and sometimes may be seen moving across the sky for hours at a time; they often show a wave motion at right angles to their length; they may show, too, traces of a rotation round their long axes (Fig. 6). These clouds are usually at heights of 20,000 to 25,000 feet. At sunrise and sunset they take up pink and rosy tints. Clouds resembling cirrocumulus are very occasionally seen at the extraordinary height of fifty miles, whereas ordinary clouds do not occur much above five miles; on a summer night such clouds appear to shine by their own light, though really it is by reflected sunlight in which they float long after the sun has set on the earth's surface. The nature of these 'noctilucent' clouds is still obscure.

A mass of cirrocumulus is often thicker in the middle than at the edges, and in such cases the edges are often rippled; the whole cloud is lens-shaped and is known as lenticular cloud (Fig. 16). The rippled edges, if closely watched, will be seen to be continually changing, and the cloud mass may be seen to grow on the windward side and to dissolve on the leeward side, so that the whole cloud remains in the same place while the actual cloud particles are moving through it.

3. Altocumulus. There is not much to distinguish very high altocumulus from cirrocumulus; but if there are no interspersed cirrus threads and if the cloudlets show any traces of shadow, the cloud in question should be classed as altocumulus. This cloud may be composed of detached cloudlets (Fig. 17), called by the French 'petits moutons' if they are small, and 'gros moutons' if they are large; the cloudlets may be arranged in waves or bands, or the cloud layer may consist of waves without any separate cloudlets (Figs. 18, 19). It is difficult to give an estimate of the height of altocumulus, for it varies greatly; possibly 10,000 to 20,000 feet would be a fair estimate. As is the case with cirrocumulus these clouds take on the colours of sunrise and sunset.

There is a curious variety of this cloud called altocumulus castellatus, or turret cloud, in which the individual cloudlets are drawn up vertically into miniature cumulus with white gleaming tops and flat grey bases (Figs. 20, 21). Turret clouds may form extensive sheets which are often very beautiful; they are an almost sure precursor of thundery conditions in the summer.

Just as a structureless sheet of cloud may exist at the cirrus level (cirrostratus), so the same formation may be found at the level of altocumulus, especially just before rainy conditions spread over the country from the west. This cloud, called altostratus, is thicker than cirrostratus; the sun may shine through it with a 'watery' look (Fig. 22), or it may obscure the sun altogether. Cirrostratus is composed of ice crystals, and halos are usually seen in it, and more rarely mock suns and other optical phenomena, but these are never seen in altostratus clouds which are composed of minute water drops.

4. Stratocumulus. This is the lowest form of sheet cloud and may occur at any level up to about 10,000 feet. Just as high altocumulus merges into cirrocumulus, so high strato-cumulus merges into altocumulus. It may show a wave or roll structure often with patches of blue sky showing between the waves. This cloud does not give the beautiful skyscapes that are seen with the higher clouds; it is on the whole a dull cloud. In quiet weather in winter it often covers the sky with a persistent cloud canopy, which may last for several days at a time (Fig. 23); sometimes it is thin and patches of blue sky may become visible, at others it is persistently thick and grey. In summer it has a tendency to break up and to become cumulus (Fig. 24); a grey morning, with the sky covered with stratocumulus, is often at this season the forerunner of a fine day.

5. Stratus. This cloud may be classed with the sheet clouds, but it is generally seen in isolated patches which are often lenticular in shape (Figs. 25, 26); these may form over and among mountains and hills, and patches of

stratus may generally be seen on the margins of large thunder-clouds (Fig. 41), appearing dark against the white cumulus. Stratus has been likened to a fog which is not in contact with the ground.

6. Fog is a cloud which is in contact with the ground. It may happen that the general cloud sheet is lower than the tops of the hills, and in such a case a fog is experienced on the hills, but this can hardly be classed as a true fog. A fog may form when damp air is chilled. This may happen in a variety of ways; for instance, in winter when a warm damp air comes in from the south over ground or sea that has been chilled by previous cold weather the current itself is cooled and a fog may be produced. Again during a cold, still and clear night the surface of the ground is cooled and so cools the air in contact with it; this air runs down slopes and fills hollows and valleys with cold air, which may become so cold that part of its water vapour is condensed into a fog. On such an evening, even before any fog has begun to form, anyone motoring or cycling along an undulating road will feel the chill of the air as he descends into hollows, and feel the air warmer again as he goes up the next rise. At about sunset on such occasions fog may be seen beginning to form in the hollows, especially over damp meadows or marshy ground. By the early morning such fogs may fill a whole valley (Figs. 27, 28). On summer mornings they soon disperse, but in winter they may last well into the day, and sometimes they may not clear away all day, and people living on the hills may be enjoying brilliant winter sunshine, while in the valleys their neighbours may be in fog so thick that traffic is interfered with. When a narrow valley runs down to the sea a valley fog may be seen flowing out to sea for a considerable distance in the early mornings; such fogs may be seen in Devon and elsewhere where narrow valleys run right down to the coast.

II. Heap clouds are clouds with a considerable vertical thickness, with rounded tops and flat bases. The smaller ones are called simple cumulus clouds (Latin *cumulus*, a heap), the larger ones that cause showers and thunder-storms are called cumulonimbus clouds. Simple cumulus is sometimes rather difficult to distinguish from stratocumulus when the latter is breaking up into cumulus, as often happens on summer mornings.

The mechanism by which cumulus clouds are formed is somewhat as follows: on a hot day air near the ground is heated, as it is heated it tends to rise, in rising it expands, and in expanding it cools; if conditions are favourable, and the current rises to a sufficient height, the cooling will be such that part of the moisture it contains condenses into a cloud (Figs. 29, 30). A cumulus cloud is therefore the visible effect of a rising current. This fact is well known to airmen, and especially to gliders who make use of the rising currents to get a lift for their sailplanes. Wherever anything tends to make the air rise there will be a tendency for cumulus clouds to form. When a wind blows on to a hilly coastline we sometimes see a line of cumulus over the coast (Fig. 31). When two wind currents converge at a small angle the air between them will be forced to rise and a long line of cumulus may be formed (Fig. 32). These lines may be very long, especially at sea. I once saw one in mid-Atlantic that lay nearly parallel to the course of the steamer and was visible all day till the late afternoon, when the steamer passed under the cloud and experienced a slight shower.

Sometimes the sky may be covered with small cumulus clouds of no great vertical thickness; they often have a tendency to be arranged in parallel bands (Fig. 2). A cumulus cloud of any size has a flat base (Fig. 33), and when many such clouds appear they will all have their bases at about the same height above the ground, though of course those that are near will appear higher in the sky than those that are far off. Such small cumulus clouds may form and never develop into large clouds; in fact on most fine summer days it is the normal thing for cumulus to form in the morning and to disappear again in the evening.

Under certain conditions, however, they grow to larger proportions (Figs. 34, 35) and may rise to towering heights (Fig. 36).

After a certain stage of growth a curious change is seen to occur at the top of a large cumulus; the hard rounded head begins to get soft and to fray out into a fibrous mass resembling some forms of cirrus (Fig. 37), and which therefore has been named false cirrus; a better name, which has been approved by the International Commission for the Study of Clouds, is hybrid cirrus (*cirrus nothus*), for the cloud is really cirrus, being composed of minute ice crystals, though it is formed in a different way from other cirrus. When this change has occurred in a cumulus cloud it is called cumulonimbus. Some rain may fall from any large cumulus, but it is only when the hybrid cirrus is developed that heavy showers and thunder-storms occur. In heavy thunder-storms there may be a great development of hybrid cirrus, and the whole top of the cloud may lose its cumulus shape (Figs. 38, 39, 40). Frequently the hybrid cirrus spreads and gives the cloud top the shape of an anvil (Fig. 38). Anvil clouds seen on the horizon are a pretty sure sign that thunder-storms are happening in the distance. Cumulus may occur at many heights; the tops of large clouds may be at 20,000 feet or more, and may be seen at great distances (Fig. 41). I have seen such cloud tops at a distance of 160 miles.

At night when a large mass of cumulonimbus is causing a thunder-storm the plumes of hybrid cirrus above the clouds may be lit up by the lightning flashes and have the appearance of flames suddenly shooting up from the clouds; such appearances may be seen a long way off; on one occasion a very intense thunder-storm occurred a few miles off the east coast of Kent, and flashes of light from the cirrus plumes were seen all over the south-east of England, as far as Winchester 120 miles to the west and Market Harborough 130 miles to the north-west.

Cumulus and cumulonimbus clouds when opposite to

the sun are gleaming white; when at right angles to the sun deep shadows are seen where rifts and valleys occur between the mountain-like peaks; when seen against the sun they are dark, and have a 'silver lining' when the sun is directly behind them.

III. The remaining form of cloud which we have not yet considered is the grey formless cloud of a wet day, known as nimbus or rain cloud. It is formless and grey when seen from below, but as seen from above it is white and often has cumulus tops projecting from the main cloud sheet; often, too, it has other forms of cloud above it, and when breaks occur in heavy rain clouds cirrocumulus and cirrus may be seen through the gaps.

Associated nearly always with nimbus, and sometimes with other clouds, are dark ragged clouds which go by the name of scud. They are usually low clouds and often appear to be moving fast, and look as though they were being torn by the wind. Scud may sometimes be seen with no other cloud in the sky, but it is generally below some other form of cloud, usually altostratus or nimbus, that it is seen.

It is interesting to watch the changes in the sky as rainy weather approaches from the west. First there may be cirrus plumes which spread out over the sky and thicken till the whole sky is covered with cirrostratus; the sun may be shining through the clouds with a 'watery' look; the cloud thickens and the sun disappears, and the clouds have developed into altostratus; then perhaps masses of scud are seen advancing below the cloud sheet, and then rain may begin to form from the cloud which has now become a nimbus. When the centre of the disturbance has passed, the sky begins to clear; this may happen very quickly sometimes. The cloud canopy breaks up into separate cumulus or cumulonimbus, from which passing showers may fall. As the depression passes still farther away the cumulus becomes less, until finally the sky clears.

This is a typical sequence of weather; but with the weather things do not always happen according to plan.

A word may be added about the clouds formed, usually at great heights, by aeroplanes, now known by the name of condensation trails (Fig. 42). These are truly cloud formations and not like the smoke trails that were once used for sky-writing advertisements.

When warm damp air is mixed with cold damp air some of the moisture will condense, and if the air is sufficiently damp and the difference of temperature sufficiently great the condensation will be sufficient to form visible cloud. The exhaust from an aeroplane engine contains a good deal of moisture, and if the air into which it is streaming is sufficiently cold and damp a condensation trail will be formed. Sometimes the trails are very slight and short lived, sometimes they are very persistent. Naturally the latter is the case when the cold air is already practically saturated with moisture, as it is when there are traces of natural cloud at the level at which the condensation trails are formed.

THE COLOUR OF THE SKY

Light consists of waves of the same nature as those used in wireless transmission. The waves used by the B.B.C. range from about 20 to 1800 metres* in length, but the waves which produce the sensation of light are of almost inconceivable minuteness, ranging from 4/10,000 to 7/10,000 of a millimetre in length, a millimetre being the thousandth part of a metre. White light consists of waves of these and all intermediate lengths. With a wireless receiver we can separate one set of waves from another; with a prism we can separate the light waves. Those measuring round about 7/10,000 of a millimetre are those which produce red light; the shorter, round about 4/10,000 of a millimetre, produce violet light; between these, in order of decreasing wavelength, come orange, yellow, green, and blue. All these

---

* A metre is rather over 39 inches, and may be thought of as a long yard.

light waves, these different colours, striking the eye at once, produce the sensation of white light.

The blue colour of the sky is caused by dust particles which exist in the air, or by the actual molecules of the air. Sunlight passing through the air has the shorter blue waves scattered by the dust particles, while the longer waves, those producing yellow and red, are less scattered. Something rather analogous can be seen in ripples and waves on water; ripples on the surface when they strike a boat are reflected in all directions, but big waves merely make the boat rise and fall, and themselves go on unimpeded. In the atmosphere there is thus a selection, the blue light is scattered in all directions and consequently blue light reaches us from all parts of the sky, while the red and yellow light goes on with little hindrance. When the sun is low its rays have to pass through a great thickness of air before they reach our eyes, and nearly all the blue rays have been filtered out; but the yellow and red rays reach us and the setting sun appears red. Soapy water, which contains a number of very fine soap particles, looks red when we look through it and blue when we look at it from the side; the same is true of tobacco smoke, which consists of very minute drops of liquid, small enough to scatter the blue light while letting the yellow and red pass through, as may be seen by blowing cigarette smoke into a paper cylinder.

SUNSET COLOURS

It has been explained why the sun looks red when near the horizon. Clouds turn pink or red at or after sunset because the sunlight which still reaches them has passed through a great thickness of the atmosphere, and has had all its blue rays filtered out and scattered, and only the red and yellow rays reach the clouds. But besides the ordinary sunset tints on the clouds there are a succession of appearances that may be seen when the sun sets in a clear sky. They may be classified as follows.

A. THE FIRST TWILIGHT

1. The first counterglow. The pink colour just before and a little after sunset, which is seen in the eastern sky, is caused by sunlight reflected by particles in the atmosphere, after it has had its blue light filtered out as explained above.

2. The first earth shadow. The sky near the eastern horizon shows a dark band below the counterglow when the sun has set; as the sun sinks lower below the western horizon this band rises higher above the eastern, and is seen to be arched; this is merely the shadow of the earth on the atmosphere.

3. The first twilight glow. This is first seen in the west as a whitish patch over the sun a little before sunset; after sunset it becomes more evident, and as the sun goes down it spreads out along the horizon, forming a bright band beneath the blue sky and the yellow horizon.

4. The first twilight arch is the upper edge of the first twilight glow.

5. The first purple light appears after the sun has set; it is a diffused purple patch, semicircular in shape and fading off at its edge till it merges imperceptibly into the blue of the sky. In our latitudes it can only be seen on exceptionally clear evenings.

These phenomena can be explained by the sunlight shining on the upper parts of the atmosphere after the sun has set on the ground.

B. THE SECOND TWILIGHT

All the above effects appear a second time, the second appearances being due not to actual sunlight, but to the sunset glow in the far west shining on the upper atmosphere, as did the sun itself during the first twilight. The second twilight effects are mostly very faint, and some of them can hardly be seen in our country except on very favourable occasions, and even then they would have to be watched for closely. In any case they are only visible when the sky is very clear.

Weather modifies the sunset very much, and sometimes, instead of the reddish or yellow twilight glow, the western sky assumes a greenish hue; this may be due to much water vapour in the atmosphere, for water vapour cuts out red light, and since the blue has been cut out by scattering, only the green gets through.

SUNSET RAYS

We often see rays of light radiating from the sunset; they occasionally extend to the zenith, and more rarely right across the sky to the eastern horizon. They are of the same nature as the rays of sunlight seen when the sun is shining through gaps in clouds, the phenomenon popularly known as the sun 'drawing up water'. When mountains or cloud heads lie on our horizon at sunset they will cast shadows across the dust-laden air above us; the parts of the sky in the shadows appear dark, the other parts light. If the clouds or mountains are extensive most of the sky may be in shadow and the remaining parts look like beams of light radiating from the horizon; they are of course parallel, and only appear to radiate from the effect of perspective. If there are only a few clouds or mountains most of the sky will be bright with dark bands radiating from the horizon. I once saw, when in the north-east of Hampshire, such a dark band extending right across the sky; it was the shadow of a cloud which was over Plynlimmon in Wales.

THE GREEN RAY

If the sun is watched as it sets on a sea horizon the very last edge of the sun to be seen is a brilliant blue-green. The appearance must be looked for very closely, as it only lasts about half a second. The explanation is that when the sun is very low it is drawn out into a very short spectrum with the red end below and the violet end at the top; but the violet and blue are cut out as explained above (p. 11) and therefore the very last part of the sun to be seen is green.

The green ray may be seen at sunrise too, but one must keep one's eye on the point where the sun is about to appear, not always an easy thing to do. Those who have seen the green ray at sea are often surprised to know that it can be seen on land when the sun sets behind a distant horizon; it may also be seen when the sun goes down behind a sharp-edged cloud very low down in the sky. The converse of the green ray is the red ray, seen when the lower edge of the sun reappears below a sharp-edged cloud which is very low down. Owing to the conditions under which it can be seen, suitable clouds in the right position, the red ray is very much rarer than the green ray.

RAINBOWS

The ordinary rainbow is sufficiently familiar, but one not infrequently finds mistakes concerning it, the most obvious one being a reversal of the colours. There are normally two rainbows to be seen, an inner bright one, and an outer faint one; the rule about the colours is that the two reds come together, that is, the inner rainbow has the red on the outside edge, and the outer bow has the red on the inside edge. The colours are of course the ordinary colours of the spectrum, though the violet end is not usually seen. Inside the primary or inner bow, and outside the secondary or outer bow, may sometimes be seen what are known as supernumerary bows, one or more according to circumstances. The colours of these are not so pure as those of the rainbow proper, those most usually apparent being pink and green. The supernumerary bows often change rapidly, fading away and appearing again at intervals. Though a typical rainbow consists of all the colours of the spectrum, it often happens that some of the colours are missing, commencing from the violet end; when for instance the sun is low down and shining through a misty atmosphere, so that all the blue light has been cut out of the sun's rays, blue cannot appear in the rainbow. When the sun shines on a very fine drizzle, which very rarely happens, or when, as happens rather more frequently, it shines on a bank of fog, a nearly white bow is seen; this is usually called a mist bow.

It should be noticed that the sky inside the primary bow is distinctly lighter than that outside.

The position of a rainbow is of course opposite to the sun, and the top of the arch is directly opposite the sun itself. The centre of the circle of the bow is the point directly opposite to the sun's position, so that in general the bow is less than a semicircle; if, however, the sun is on the horizon the bow is a semicircle, but this does not often happen. The higher the sun in the heavens the smaller will be the segment of the bow that is visible, and in summer in England the sun is so high that, in the middle of the day, no part of the rainbow can be seen against the sky, but sometimes when the sun is high and there is rain close at hand and very bright sunshine, a small segment of the bow may be seen projected against the background of trees or hills; this would be specially noticeable if the observer, standing on a hill, were looking down on to a valley where rain was falling.

It may be useful to give the actual angular dimensions of a rainbow. From the centre of the circle, that is, from the point directly opposite to the sun, which point will in general be below the level of the horizon line, it is 41 degrees to the middle of the primary bow and 53 degrees to the middle of the secondary bow.

A lunar rainbow is very rarely seen; it can only be seen when the moon is getting on for full, and therefore the number of nights on which it can be seen is limited, and heavy showers with patches of clear sky between, the conditions necessary for a rainbow, are much rarer at night than during the day. Owing to the faintness of moonlight compared to sunlight, the colours are not usually seen, just as the colours of flowers are not usually seen by moonlight; a lunar rainbow is therefore practically white. Naturally the position of a lunar rainbow with regard to the moon follows the same laws as those of the ordinary rainbow with regard to the sun.

Occasionally portions of rainbows are seen which are not in the normal positions; for instance, an arc of a bow is sometimes seen extending upwards and outwards from the base of an ordinary rainbow. This is probably caused by the reflection of the light of the sun from still water. These abnormal bows however are very rare.

HALOS

Besides rainbows there are other coloured rings that are of a different nature; these are called halos and are seen round the sun or moon. The halo most commonly seen is a ring 22 degrees from the sun or moon; it is usually whitish, but sometimes colours are visible; in this case red is on the inside of the circle, followed by yellow and green; when seen round the moon the colours are hardly visible. A halo is an optical phenomenon caused by the sun or moon shining through clouds composed of minute ice crystals, but sometimes these clouds are so thin that the only indication of their presence is the fact that a halo is visible; it is on these occasions that the colours are best developed. An outer halo 46 degrees from the sun or moon is sometimes seen. There is sometimes a white circle going right round the sky through the sun and parallel with the horizon; it is sometimes called the mock sun ring, for on it are situated the mock suns, or parhelia; the two most generally seen occur close to the points where the mock sun ring cuts the 22-degree halo; they are just outside the halo, and the higher the sun in the sky the farther off are they from the halo; they are sometimes white, but generally show prismatic colours with the red nearest to the sun; they are more or less drawn out on the side away from the sun; they may be seen on many occasions when no halo is visible; sailors call them sun dogs, and they are popularly supposed to be a sign of bad weather, but this is by no means always true.

There are other mock suns on the mock sun ring, one exactly opposite the real sun, but they are very rare. Sometimes small arcs of circles may be seen touching the other halos above or below the sun. Another arc of a circle is the circumzenithal halo; only about a quarter of the circle is visible, the part that lies toward the sun; being very high up and nearly overhead it is often missed, and it is commoner than one would suppose; the colours are often brilliant and are as pure as those of the rainbow. It is a very striking phenomenon.

There is a very great variety of other optical phenomena connected with halos, but they are mostly very rare.

Fairly frequently a sun pillar may be seen extending vertically upward from the sun at about the time of sunset; this is usually white, but it may be reddish, especially just after sunset. This appearance is also caused by minute ice crystals in the upper air; the sunlight is reflected from their faces, and the phenomenon is somewhat analogous to the band of light, or 'wake', seen in the sea or other sheet of water under the sun or moon.

CORONAE

These are coloured rings round the sun or moon, but they are quite different from halos. They are quite close up to the sun or moon and are formed when these are shining through light clouds composed of minute drops of water. The colours are in the reverse order to the halo, the red being on the outside; sometimes the colours are repeated one or more times, in all cases the red being the outer colour of each circle. Since halos are formed by clouds composed of ice crystals, and coronae by those formed of water drops, we do not get the two together as a general rule; but it occasionally happens that a halo and a corona are seen at one and the same time. This may happen when there are two layers of cloud, one composed of ice crystals, the other of water drops, both being sufficiently thin for the sun or moon to shine through both. Coronae are not often noticed round the sun as they are lost in the glare of the sunlight; they may very often be seen with the aid of dark glasses, or reflected in pools of water, when they cannot be seen by direct vision.

IRIDESCENT CLOUDS

Sometimes light clouds of the higher varieties are seen to have colours developed in them; red and green are the most prominent; they are seen in clouds near the sun, but there is no definite distance from the sun at which they make their appearance, and the colours are arranged in an irregular manner; they are not nearly as rare as might be supposed, but they are often missed, as are coronae, owing to the glare of the sun; like coronae they may be seen by the help of dark glasses or reflected in still water.

BROCKEN SPECTRES

These are apparently huge shadows seen on mist. They were observed on the Brocken in Germany, hence their name, but they may be seen under suitable conditions in any hilly country. I have seen one on the Portsmouth Road about two miles south of Butser Hill. About a couple of hundred yards away the ground fell away into a valley that was filled with a slight mist; beyond, the ground rose again at a distance of between a quarter and half a mile; my shadow was projected on the mist about 200 yards away, but it looked as though it were on the opposite slope nearly half a mile away. That is where the mind's eye placed it, and so it appeared to be inordinately large.

Under suitable conditions a coloured ring may be seen round the shadow of one's head on a mist; such a ring is called a glory. If there are several observers each will see a glory round the shadow of his own head but none round those of his neighbours. A glory may also be seen by airmen round the shadow of their own plane on the clouds.

MIRAGES

A mirage is a subject about which there is a good deal of misconception. A flat surface, when heated by the sun, heats in its turn the layer of air that is in contact with it. When a ray of light falls on a heated layer of air at a glancing angle it is bent upward again as though reflected. If an observer stands at A and looks at a point R on, say, a heated road surface, he will see objects which may lie on the line RS; if there are no solid objects on this line he will see the sky at S apparently reflected at R, and the road

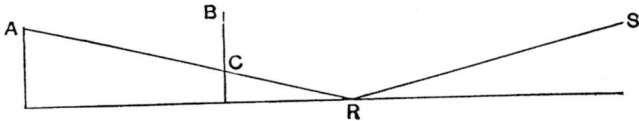

surface at R will appear as though it were covered with water. If the observer goes nearer to R, say to the point B, he will no longer see the mirage at R because no ray of light falling on the road surface at R can be bent up sufficiently; it is only glancing rays that are bent up and they are not bent up except at very small angles. If, however, the observer at B stoops down so that his head is in the position C, he will see the mirage again as though he were at A. I have been told by one who had motored in the desert that he once saw a mirage in every direction, so that he seemed to be on an island that was surrounded by a sea; but his island moved with him; he naturally could not see the mirage close to him for the reason explained above. The writer who described a desert scene when the mirage was so close that a rider saw his camel nearly putting his foot into it was describing an impossible phenomenon.

Mirages are not only seen in the desert. In the summer they can be seen on roads in England, especially on tarred roads which become intensely heated by the sun. The best place to see them is just before one comes to the top of a rise, when the observer's eye looks along a level stretch ahead; quite a good one is to be seen in this way on the road approach to Waterloo Station.

THE MOON

The moon is not connected with meteorology but it is what may be called a sky phenomenon, and it may not be out of place to mention here a few facts about it, as mistakes are often made, judging by the works of landscape painters.

Since the light of the moon is merely reflected sunlight it follows that the convex side of the crescent or of the half moon is always turned towards the sun's position. The direction in which the half moon is seen must be at right angles to the sun's position whether the sun is above the horizon or below it; the crescent moon must be nearer the sun than a right angle; the smaller the crescent the nearer it is to the sun and therefore a very thin crescent must not be depicted a long way from the sunrise or sunset; the new crescent moon is near the position where the sun has set, the old 'decrescent' moon is near the place where the sun will shortly rise. A full moon rises in the east at the time the sun is setting in the west, it is south and high in the sky at midnight, and it sets at sunrise. A gibbous moon, that is, a moon that is between half and full, rises before sunset when the moon is waxing, and after sunset when the moon is past full or waning. A gibbous moon has its sharp convex edge directed to the sun's position, and therefore it has its sharp edge turned to the west before and to the east after full moon. This point is not always noticed, and I have seen a picture whose title indicated that evening was in question, with the gibbous moon rising and its sharp edge turned to the eastern horizon, an impossible position.

The position of the crescent moon calls for further note. In the spring the crescent moon is nearly above the sunset, though a little to the left, and the crescent is 'on its back'. This is a purely astronomical condition and has nothing to do with the weather; the March crescent is always on its back. As the year advances the crescent moon moves, at each new moon, farther to the south, or to the left of the sunset, and is lower down on the horizon; by the autumn the new crescent moon is low down to the left of the sunset, and is very nearly upright; as the season advances further the crescent, at each new moon, gradually moves back till it reaches its March position again. Under no circumstances can the crescent moon be seen to the north or right-hand side of the sunset in our latitudes. In the southern hemisphere however things are reversed; the crescent moon will be on the north or right-hand side of the sunset position; the March crescent will be low down to the north of the sunset, the September crescent will be high over it.

At sunrise the old decrescent moon will, in the northern hemisphere, be low down to the south or right of the sunrise in the spring, and will be nearly upright; in autumn it will be high up over the sunrise, and on its back in middle northern latitudes. In the southern hemisphere things will be reversed, as with the new crescent. In the tropics the new crescent and the old decrescent will be seen on its back over the sunset and sunrise respectively all through the year.

The full moon is high up in the heavens at midnight in winter, and low down in the summer; in both cases it is toward the south. It is a rough rule that the full moon travels across the sky in nearly the same path that the sun took six months before.

Contrary to popular opinion there is no connection between the phases of the moon and the weather. The matter has been thoroughly investigated by meteorologists, and if any connection had existed it would certainly have appeared. The only effect on the atmosphere that has come to light are minute changes of the barometer which are far too small to have any effect on the weather, and it has taken elaborate mathematical analysis to bring them to light.

THE APPARENT SIZE OF THE SUN & MOON
ON THE HORIZON

The sun and moon, and for the matter of that the constellations also, appear to be much larger when near the horizon than they do when high up in the sky. The effect is not a real one; it is perhaps due to our estimation of the shape of the sky. We do not estimate this as a hemisphere but as a shallow inverted bowl; consequently we imagine the sky at the zenith to be nearer than that at the horizon. The sun and moon have the same angular diameters whether they are high up or low down, but they appear larger when on the horizon because we are imagining them to be farther away; if you see a man one hundred yards away, but for some reason or other estimate his distance as two hundred yards, he will appear to be a giant. Hills and mountains on the horizon look higher than they really are; the mind's eye increases their height, but a camera shows them in their true proportions, which is the reason why photographs which include hills or mountains are sometimes so disappointing.

I once saw an effect that I have never seen mentioned; in a very sheltered inlet of the sea I looked down from a boat on a perfectly unruffled surface; every star down to the third or fourth magnitude was reflected; the effect was the converse of the apparent enlargement near the horizon; the mind's eye put the constellations quite close, on the surface of the water, and they appeared extremely small.

INDEX

Fig. 2. Clouds at two levels, cirrus above, cumulus below

Fig. 3. Cirrus in isolated wisps

Fig. 4. Tufts of cirrus

Fig. 5. A large mass of cirrus

Fig. 6. Part of a long band of cirrus, showing signs of rotation about its axi

Fig. 7. Cirrus with upturned ends

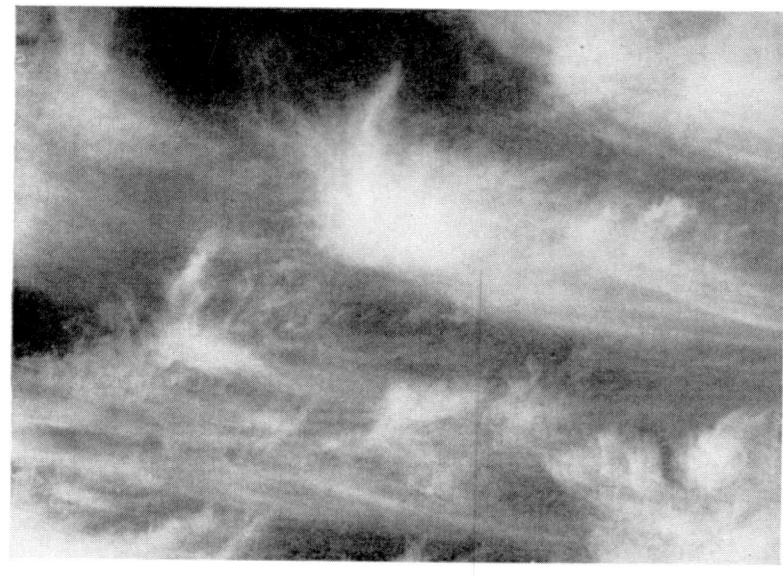

Fig. 8. Cirrus with upturned ends

Fig. 9. Tailed cirrus

Fig. 10. Cirrus plumes

Fig. 11. Bank of cirrostratus, some isolated bands of
cirrus, and some cumulus below

Fig. 12. Cirrocumulus waves at sunrise

Fig. 13. Cirrocumulus cloudlets arranged in waves and ripples

Fig. 14. Cirrocumulus band with wave structure

Fig. 15. Cirrocumulus band with fine ripple structure

Fig. 16. Lenticular cirrocumulus, and cumulus at a lower level

Fig. 17. Altocumulus cloudlets

Fig. 18. Altocumulus waves

Fig. 19. Altocumulus waves

Fig. 20. Turret cloud, slightly developed

Fig. 21. Turret cloud, strongly developed

Fig. 22. Altostratus, with altocumulus bands at a lower level

Fig. 23. Stratocumulus, quiet weather type

Fig. 24. Stratocumulus tending to become cumulus

Fig. 25. Stratus

Fig. 26. Stratus (dark) and high cloud over the south coast

Fig. 27. Early morning fog at the western end of the Weald
(infra-red photograph)

Fig. 28. Early morning valley fog; note the position of a train in a station

Fig. 29. Small cumulus clouds of fine weather

Fig. 30. Small cumulus clouds of fine weather

Fig. 31. Cumulus clouds forming over the Dorset coast

Fig. 32. Long line of cumulus, Outer Hebrides, the
result of converging winds

Fig. 33. Cumulus, showing flat base

Fig. 34. Cumulus clouds, some of them with a growing tendency

Fig. 35. Cumulus clouds with a growing tendency

Fig. 36. Cumulus that has grown to a great height

Fig. 37. Cumulus with hybrid cirrus just beginning to form at the top

Fig. 38. Cumulonimbus whose whole top has become
hybrid cirrus (Anvil cloud)

Fig. 39. A shower cloud in front of the sun with 'silver lining'

Fig. 40. A distant thunder-cloud

Fig. 41. A very intense thunder-cloud about 70 miles
distant; dark stratus in front

Fig. 42. Condensation trails